Burrows

Julie Murray

Abdo Kids Junior
is an Imprint of Abdo Kids
abdobooks.com

Abdo
ANIMAL HOMES
Kids

abdobooks.com

Published by Abdo Kids, a division of ABDO, P.O. Box 398166, Minneapolis, Minnesota 55439.

Printed in the United States of America, North Mankato, Minnesota.

052019

092019

Photo Credits: Alamy, iStock, Science Source, Shutterstock

Production Contributors: Teddy Borth, Jennie Forsberg, Grace Hansen

Design Contributors: Christina Doffing, Candice Keimig, Dorothy Toth

Library of Congress Control Number: 2018963324

Publisher's Cataloging-in-Publication Data

Names: Murray, Julie, author.

Title: Burrows / by Julie Murray.

Description: Minneapolis, Minnesota : Abdo Kids, 2020 | Series: Animal homes | Includes online resources and index.

Identifiers: ISBN 9781532185212 (lib. bdg.) | ISBN 9781644941188 (pbk.) | ISBN 9781532186196 (ebook) | ISBN 9781532186684 (Read-to-me ebook)

Subjects: LCSH: Animal housing--Juvenile literature. | Lairs (Animal habitations)--Juvenile literature. | Burrowing animals--Juvenile literature. | Animals--Habitations--Juvenile literature.

Classification: DDC 591.564--dc23

Table of Contents

Burrows

Many animals live in burrows.

A burrow is a hole or tunnel.

It is dug by an animal.

It can be in the dirt or sand. It can also be in wood or rocks.

Some are just one hole. Others have many **tunnels**.

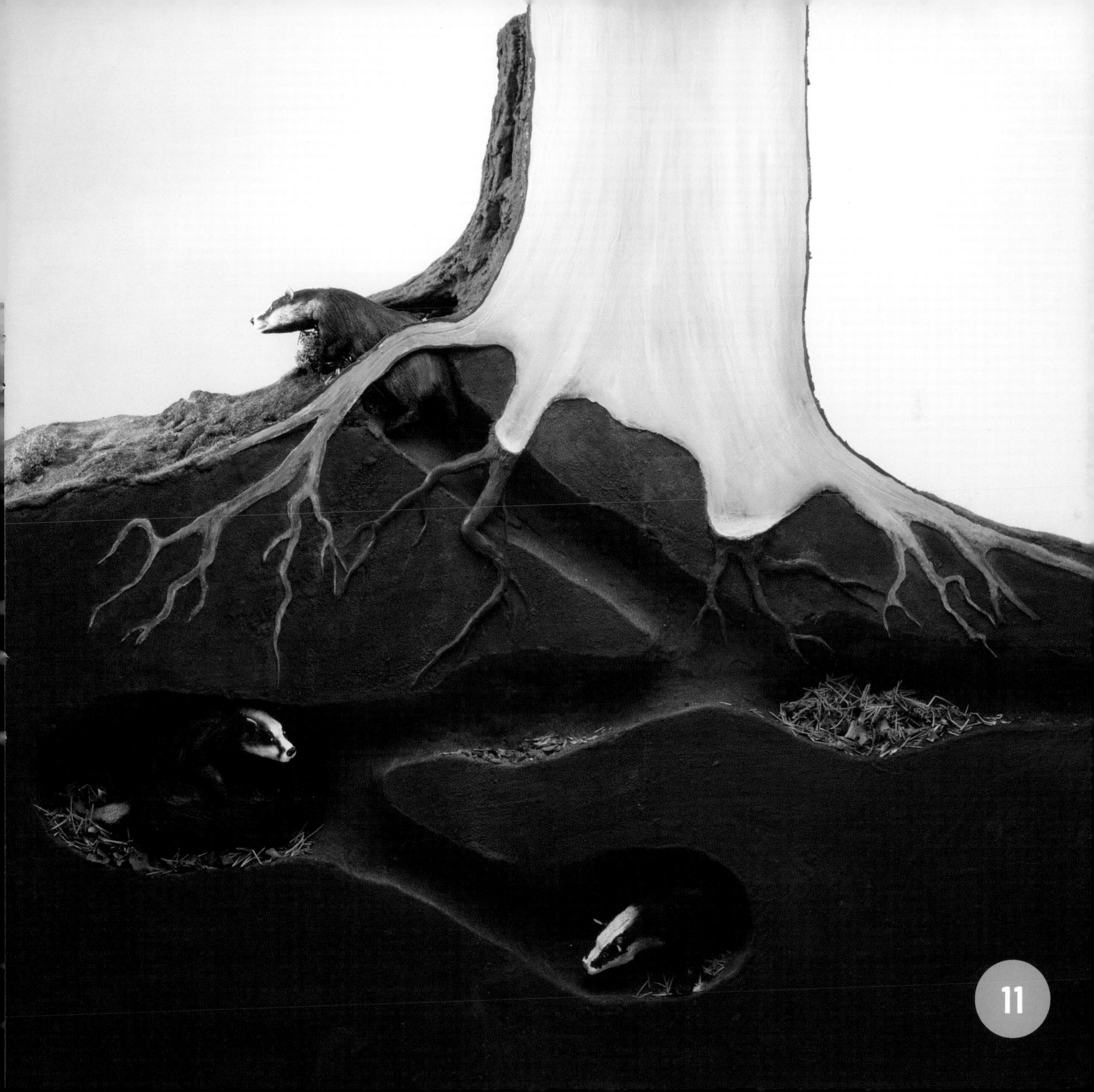

Animals hide here. It keeps them safe.

Rabbits can live in a burrow.

Lots of them live together.

Some owls live in one.

Their eggs are safe here.

The sun is hot. A Gila monster digs a hole. This keeps it cool.

A mole lives in a burrow. It has sharp claws. It can dig fast!

What Lives in a Burrow?

chipmunks

red ants

shrews

tarantulas

Glossary

dug
to have made a hole by removing dirt or sand in one area.

tunnel
the hole or den or passage between dens of a burrowing animal.

Index

Visit abdokids.com
to access crafts, games,
videos, and more!

Use Abdo Kids code
ABK5212
or scan this QR code!